Grasiele Dick
Mauro Valdir Schumacher

Soils and eucalyptus forestry in Rio Grande do Sul

Grasiele Dick
Mauro Valdir Schumacher

Soils and eucalyptus forestry in Rio Grande do Sul

Case studies

ScienciaScripts

Imprint
Any brand names and product names mentioned in this book are subject to trademark, brand or patent protection and are trademarks or registered trademarks of their respective holders. The use of brand names, product names, common names, trade names, product descriptions etc. even without a particular marking in this work is in no way to be construed to mean that such names may be regarded as unrestricted in respect of trademark and brand protection legislation and could thus be used by anyone.

Cover image: www.ingimage.com

This book is a translation from the original published under ISBN 978-613-9-64005-8.

Publisher:
Sciencia Scripts
is a trademark of
Dodo Books Indian Ocean Ltd. and OmniScriptum S.R.L publishing group

120 High Road, East Finchley, London, N2 9ED, United Kingdom
Str. Armeneasca 28/1, office 1, Chisinau MD-2012, Republic of Moldova, Europe
Printed at: see last page
ISBN: 978-620-7-72147-4

Presentation

This book presents reports and recommendations on the cultivation of trees - silviculture - in the state of Rio Grande do Sul, mainly using eucalyptus, a fast-growing exotic species, focusing on practices carried out in the campanha region of the Pampa biome. Practical, operational and ecological aspects of growing eucalyptus in the three main soil conditions that occur widely in the state, including Neossolo, Cambissolo and Argissolo, will be discussed. The main objectives of this work are to recommend planting, management and silvicultural management practices, in a conservationist and environmentally correct manner, in each of these soil conditions, as well as presenting information on the productivity observed in the state's forest plantations.

The experiences reported here come from the monitoring of practices carried out with forestry companies, which cultivate these areas with soils of low natural fertility, degraded by livestock farming and intensive agriculture, with a predominance of sand in the textural composition, among other conditions, which have induced the conversion of land use in areas of the Pampa biome to forestry. Happy reading!

The authors

Summary

CHAPTER 1
Current Forestry Scenario

The global assessment of forest resources indicated that 290.4 million hectares of the world are occupied by forest plantations (FAO, 2016), 20 million of which are eucalyptus plantations (BOOTH, 2013). Intensively managed plantations account for 1.5 per cent of the world's forested areas, and those used for eucalyptus forestry are among the most productive, although they only meet a third of the demand for wood products (BINKLEY et al., 2017).

In Brazil, the establishment of forest masses, mainly with fast-growing species, has been taking place since the beginning of the last century (ABRAF, 2012). By 2016, the area of planted trees totalled 7.84 million hectares, an increase of 0.5% on 2015, due exclusively to the expansion of eucalyptus plantations (Figure 1) (IBÁ, 2017).

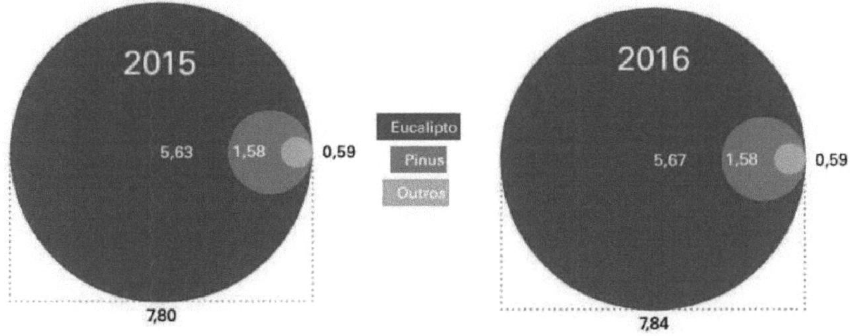

Figura 1 - Expansion of Brazilian forestry conditioned by the increase in the area planted with eucalyptus. Source: IBÁ, 2017.

Eucalyptus plantations occupy 5.7 million hectares, which represents 71.9 per cent of the total area with planted trees and are mainly located in the states of Minas Gerais (24 per cent), São Paulo (17 per cent) and Mato Grosso do Sul (15 per cent), which have been standing out on the country's forestry scene (Figure 2) (IBÁ, 2017).

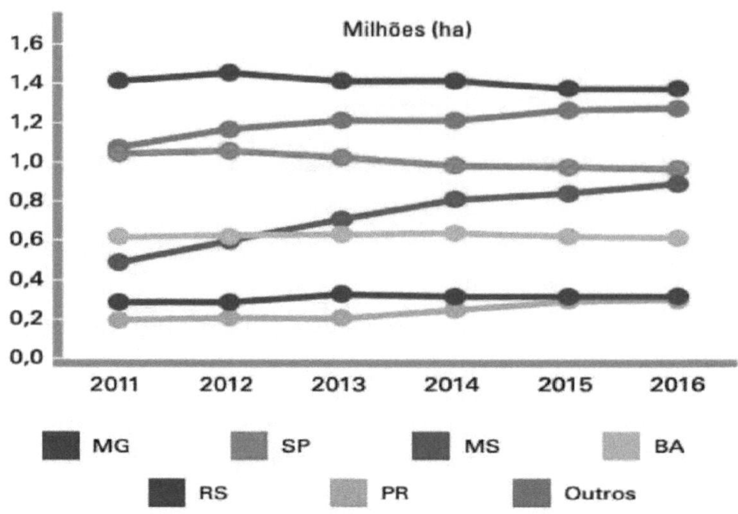

Figura 2 - Evolution of eucalyptus cultivated areas in Brazilian states. Source: IBÁ, 2017

In Rio Grande do Sul, the area with forest plantations is 596,700 ha, which is equivalent to 8% of the planted area in Brazil and 2% of the state's territory (28 million ha). The largest area is occupied by eucalyptus, which accounts for 52 per cent (310,200 ha), while pine and acacia account for 31 per cent and 17 per cent respectively. The state occupies an important place on the Brazilian economic scene, especially in terms of the significance of its forest-based production chain. In 2014, the sector contributed 4% of RS's GDP, 7% of job creation, 3% of tax collection and 2% of the sales value of exported forest-based products (AGEFLOR, 2015).

Given the importance of the sector and the expansion of forestry production around the world, in Brazil and in Rio Grande do Sul, especially using eucalyptus species, it is necessary to improve silvicultural processes, especially with regard to plantation productivity. This can be achieved through the selection of adapted genetic materials and the implementation of appropriate silvicultural management techniques and the addition of fertilisers (GONÇALVES et al.,

2015).

Along with this promising forestry scenario, in which higher yields are sought from plantations, the search for conservation of natural resources through investments in conservation practices, such as minimum soil cultivation, fertilisation management and planning of machine traffic and harvesting, are essential, because with the intensification of monoculture, soils are prone to nutritional exhaustion (PAES et al., 2013).

In Rio Grande do Sul, forestry generally takes place in sloping areas and/or areas with low production potential for agricultural crops and pastures. In recent years, forestry frontiers have been expanding, mainly in the Pampa biome, due to the need to supply raw materials to the consumer market, business development and investment, low land costs, among other motivations.

The success of establishing forestry in the Pampa reflects the application of appropriate treatments, as well as the choice of species, taking into account physiological and productive aspects depending on each site and climatic conditions (BOOTH et al., 2015). According to Boldrini et al. (2010), it is in the Pampa biome region that the largest expanses of continuous native Brazilian grassland can be found, however, highly impactful activities such as intensive livestock farming and conventional agriculture have weakened these areas. Currently, more than 33 per cent of native vegetation is degraded and/or in the process of being degraded (VERDUM et al., 2014).

In this context, forestry can be a viable activity for the productive utilisation of these areas, as well as helping to improve soil quality and contain erosion, since species of the Eucalyptus genus, for example, adapt to the most diverse environments (BOOTH et al., 2015).

Among the potential genera for forestry production in the state of Rio Grande do Sul, Acacia, Pinus and especially Eucalyptus stand out, as many of the species have ecophysiological plasticity and cold tolerance, and are especially destined for the pulp industry and charcoal production (FLORES et al., 2016; GLENCROSS et al., 2014; AGNELI, 2005).

In connection with the expansion of forestry in the Pampa biome, recurring questions have arisen from both research and operational forestry. What is the best approach to soil preparation, planting and fertilisation, with a view to practising the activity in a conservationist manner that reduces impacts on the soil and diversity?

Growing trees in different types of soil requires differentiated management, respecting the particularities and limitations of the soil, always aiming to maintain the productive potential of the next rotations by conserving fertility. Soil quality is directly related to conservation management in agriculture, livestock farming and forestry.

CHAPTER 2

Soils and Forestry

Forestry is practised in the most varied soil conditions, especially on low-fertility soils. This book will look at silvicultural practices carried out on three different soil classes (Figure 3): Neossolo, characterised by being a "young" type of soil, poorly developed and weathered, and can be shallow or deep; Cambissolo, more developed than Neossolo, but identified mainly by the presence of fragments of original material, apparent due to the incipient process of formation, varying in depth and drainage levels and Argissolo, which has a clayier subsurface horizon and textural gradient, more weathered and developed in relation to the previous classes, with greater depth and is the most common type of soil in Rio Grande do Sul and Brazil (STRECK et al., 2008).

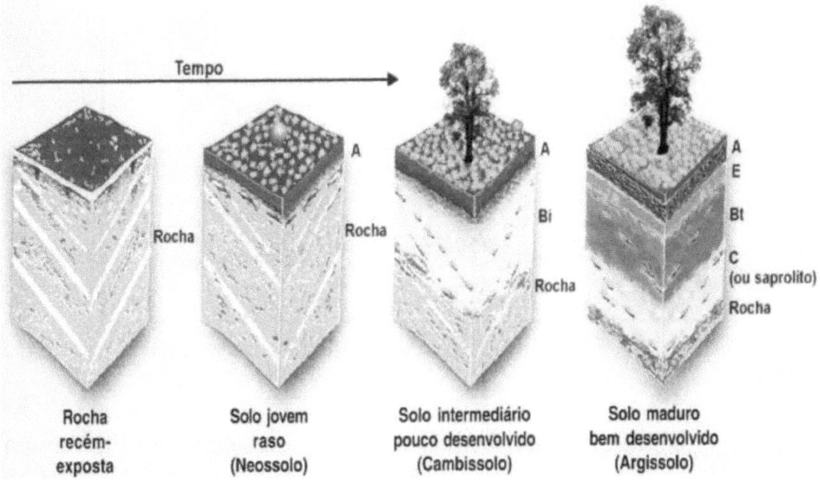

Figura 3   - Morphological and evolutionary distinctions between the Neossolo, Cambissolo and Argissolo classes. Source: Lepsch, 2002

According to the "Status of World Soil Resources" report (FAO, 2015), soils are fundamental to life on Earth, but human pressures on this resource are reaching critical limits. The problem involves drastic changes in soil quality, since every year intensive agricultural and forestry cultivation causes the loss of 5 Gt of soil through erosion. To this end, conservation soil management is advocated, which consists of carrying out sustainable practices, such as minimum tillage, which provide valuable leverage for maintaining productive capacity and biodiversity.

The state of Rio Grande do Sul is characterised by pedological diversity, which is due to the different materials of origin, climate variability, landforms and the variety of edaphic organisms

(STRECK et al., 2008). The different configurations of Rio Grande do Sul's landscapes and four of the twelve soil classes found in the state are shown in Figure 4.

Figura 4   - Pedological diversity and landscapes in the state of Rio Grande do Sul. Source: Dick, 2013.

Despite pedogenetic diversity, soil degradation, especially in the Pampa biome region, is being caused by the development of production processes such as livestock farming and mechanised and intensive agriculture. These factors, together with the effects of wind erosion (Figure 5) and water erosion (Figure 6), have led to the emergence and intensification of important landscape-modifying phenomena in the Pampa biome: gullies and the expansion of sandy cores (Figure 7) (VERDUM et al., 2014).

Figura 5   - a) Wind erosion in sandy soil; b) Intense movement of soil particles due to wind action, Maçambará, RS. Source: Dick, 2015.

Figura 6   - Erosion and soil movement caused by the action of water on a bare slope. Maçambará, RS. Source: Dick, 2015.

Figura 7   - Cracking and arenisation phenomena in the western campaign region of the Pampa biome, Maçambará, RS. Source: Schumacher, 2015.

The practice of forestry in these areas subject to degradation, such as the physiographic region of the Pampa campaign, requires careful planning in terms of minimum cultivation and waste management to avoid leaching and depletion of soil nutrients (GONÇALVES et al., 2004). Specifically for eucalyptus, the soils used for planting eucalyptus are generally of low fertility, while the forest management techniques used to grow the species in Brazil are still very intensive (BARROS et al., 2014).

In this context, knowledge about the effects of forestry on soil nutrient reserves and balance, especially in short rotation plantations, is essential for defining conservation forestry practices (LEITE et al., 2010). In the long term, the sustainability and productive capacity of these ecosystems is probably conditioned by the maintenance of organic matter and nutrient availability, especially in areas where the soils are of low fertility (BELDINI et al., 2009). For example, Laclau et al. (2010) found a one-third increase in the volume of wood in eucalyptus clones when organic matter was conserved through the maintenance of residues, and consequently there was an increase in nutrient reserves in sandy soils.

Changes in soil nutrient balances (Figure 8) can affect all types of vegetation, however, these are more likely to occur in forest areas where biomass is exported. Monitoring soil nutrient stocks is highly relevant in forest plantations, as it is a means of assessing their ability to maintain ecosystem functions (FAO, 2015).

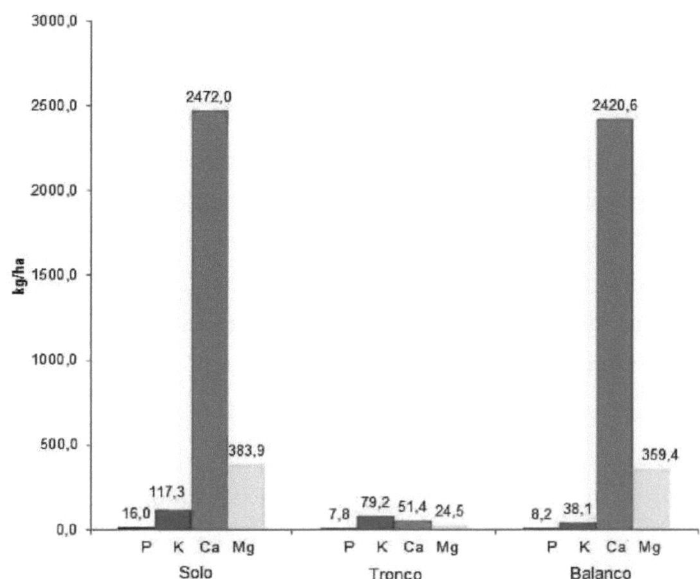

Figura 8   - Example of results from analysing the nutrient balance in the soil-plant system, 5 years after planting. Study carried out on a Eucalyptus dunnii stand grown on Argissolo in the municipality of Alegrete, RS. Source: Dick et al., 2015.

In forestry plantations, where trees are harvested at the end of the rotation, there can be major changes to the soil, due to the high degree of anthropogenic disturbance, which makes it essential to know the characteristics and rational use of the soil, seeking conservation combined with improved production potential (GONÇALVES, 2002). To this end, forestry requires efficient techniques and management that optimise production while conserving natural resources, thus maintaining soil quality.

The soil is responsible for providing essential nutrients to plants, which are supplied through the application of fertilisers, inputs via rainfall, rock weathering, decomposition of waste, adsorbed on colloids and released into solution for subsequent root absorption (TAIZ and ZEIGER, 2013; MEURER et al., 2010; BEGON et al., 2007). The morphological, physical, chemical and biological characteristics of the soil can also interfere with the availability of nutrients by retaining them with greater or lesser energy, influencing root growth,

11

providing nutrient transport towards the root surface, among other factors (BARROS et al., 2014).

The changes to the soil characteristics in the different soil classes caused by the practice of forestry, as well as the consequences of these changes on the sustainability of the eucalyptus production process, in particular, are still little known (LEITE et al., 2010). According to Lima (1996), studies involving the chemical properties of soil in forest stands, aimed at detecting changes in pH, organic matter content, nutrient stocks, cation exchange capacity, among other aspects, can indicate the effect of eucalyptus plantations on soil quality.

CHAPTER 3

Silviculture on neosols

The case study on the practice of forestry in Neossolo was carried out in a eucalyptus plantation, accompanying the planting in a sandy core, in the municipality of Maçambará, in the western campaign region of the Pampa biome in Rio Grande do Sul (Figure 9). The idealisation of forestry on such an unfavourable site required a great deal of planning, observations of previous experiences, as well as a joint effort between the knowledge produced at the university and the mistakes and successes made by companies in the forestry sector.

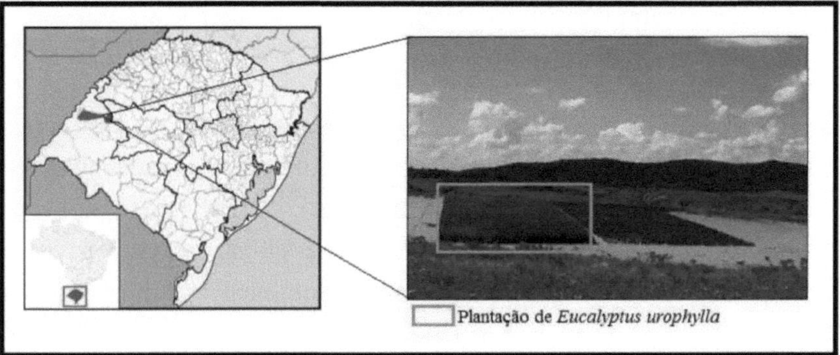

Figura 9   - Location of eucalyptus plantation on Neossolo, Maçambará, RS. Source: Wikipedia, 2018; Souza, 2017.

As well as being a complex challenge, cultivating and containing the expansion of sandy cores (Figure 10) has often been practised without great results in terms of productivity, economic return and social functionality. These areas have become underutilised within the productive context of the biome, receiving little attention from farmers and ranchers, mainly due to their low fertility and compromised productive potential.

Figura 10 - Landscape and sandy cores in the western campaign region of the Pampa biome, Maçambará, RS. Source: Dick, 2015

Sandy is a natural process, however, there is an expansion of sandy cores, which are unconsolidated sandstone deposits, devoid of vegetation and reworked under the characteristic processes of the current climate, suffering great influence from rain and wind, thus making it difficult to establish vegetation cover (SUERTEGARAY, 2011). Until 1989, the aforementioned author states that sandy areas totalled 3,024.37 ha in Rio Grande do Sul. By 2005, this area had risen to 3,027.41 ha, with 137 new sandy spots appearing during this period.

With regard to the sandy soils of the Pampa biome, despite numerous efforts aimed at recovering the edaphic quality and reinserting these areas into the regional productive and economic context (SUERTEGARAY, 2011), there is still a significant gap in knowledge about the effectiveness of forestry as a tool in this process.

In order to realise and make forestry viable on this site, it was assumed that water is not the most limiting production factor, as rainfall is well distributed throughout the year, with adequate volumes of precipitation (Figure 11) and, for this reason, using the term desert to describe these areas is wrong. In the Maçambará region, the average annual rainfall is 1,628 mm and the average annual temperature is 20.7 °C, while the average coldest month is 15.5 °C and the average hottest month is 26.3 °C. In winter there are freezing temperatures. The winds predominate from the south-west in winter and the north-east in spring (ALVARES et al., 2014).

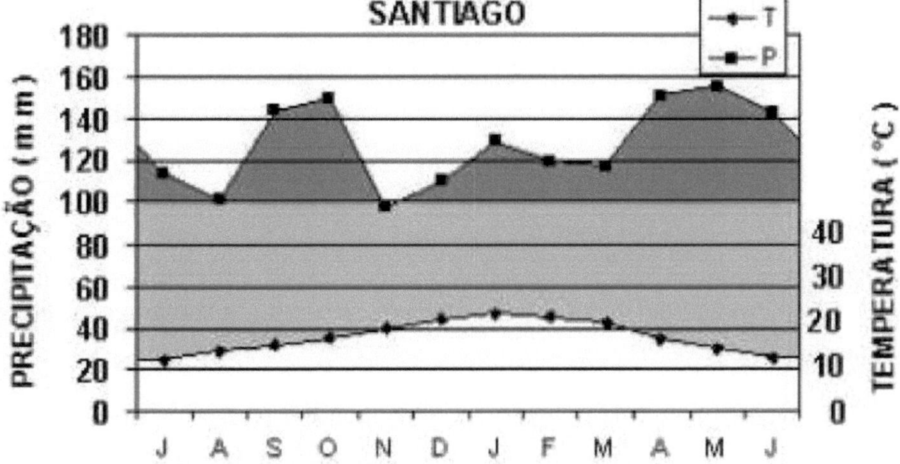

Figura 11   - Climate diagram for the municipality of Santiago, located 70 km from Maçambará, RS. Source: Buriol et al., 2007.

Another significant factor affecting forestry production is soil fertility (nutrients and soil), so, given the predominantly sandy soil class and textural composition, this was the bias defined as a guideline for forestry practice. In other words, among the possible production factors (Figure 12), the premise of managing fertilisation by dividing up the doses was considered a promising alternative; however, before adding any minerals to the soil, it is necessary to know its fertility levels.

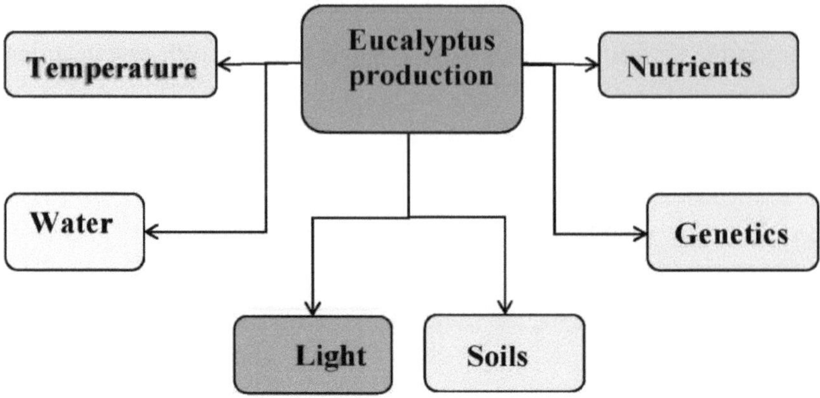

Figura 12 - Production factors that influence eucalyptus plantations. Source: Schumacher, 2018.

Previously, one of the first steps in implementing forestry in this area was to open a trench up to 2 m deep in the sandy core, from where the morphological classification was carried out, soil samples were collected for chemical analysis of nutrients (TEDESCO et al., 1995) and samples for density analysis, using the volumetric ring method (EMBRAPA, 1997) (Figure 13).

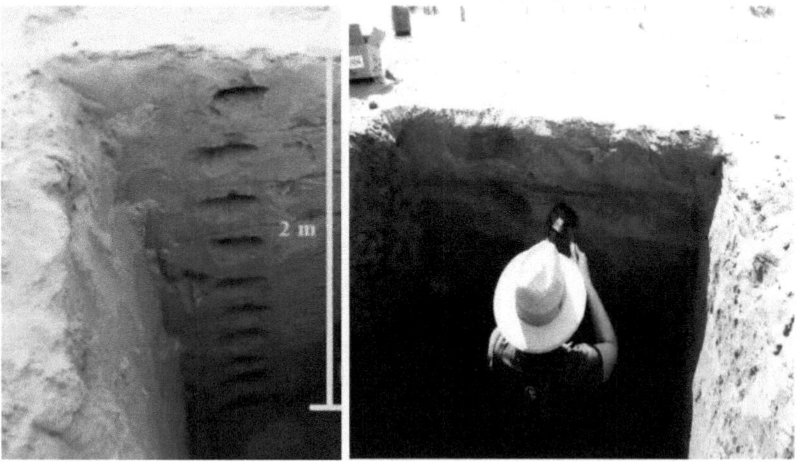

Figura 13 - Neosol profile and sample collection. Source: Dick, 2015.

As greater collection depths were reached, an increase in the soil moisture gradient could be observed, however, no distinctions were observed between pedogenetic horizons, i.e. the textural composition is homogeneous throughout the profile of this Neossolo, originating from the geological formation of the Botucatu sandstone

The soil in this sandy core, at a great depth, has been classified as typical Arctic Quartzarenic Neosol (EMBRAPA, 2013), with more than 80% coarse sand in the textural composition, low fertility characterised by organic matter contents of less than 0.3%, base saturation below 8% and acid pH up to 2.0 m deep. Fertilisation on this site is essential due to the low levels of minerals in the soil, a condition that is linked to the predominance of sand in the soil composition and the low organic matter content, which causes intense leaching, and as a result there is less adsorption and

retention of nutrients in colloids (SILVA and MENDONÇA, 2007; MEURER et al., 2010).

Another production factor that can be managed is the choice of species (genetics). In this case, the choice was made to grow Eucalyptus urophylla, based on experience and observations from previous plantations, where various genetic materials were planted in sandy soil and this species showed greater adaptation, resistance and productivity. It should be emphasised that these inferences were made based on phenotypic analysis of individuals, which does not reflect the genotypic capacity for greater efficiency in the use of nutrients and water. The literature lacks information and recommendations on these aspects.

After analysing the soil and choosing the species, the next steps and procedures carried out to establish the eucalyptus plantation in Neossolo consisted of leaf-cutting ant control, soil preparation, liming and planting fertilisation, planting the seedlings and top dressing fertilisation.

The control of leaf-cutting ants, the main pest in forestry (Figure 14), was carried out around the sandy core, in the surrounding native field, since no anthills were found at the planting site. For this stage, bulk formicide baits were used, with the active ingredient sulphuramide, applied 30 days before planting.

Due to the licensing requirements for forestry, in order to reduce environmental impacts, ant control is carried out manually, where operators identify the anthills and apply the formicide bait at a minimum distance of 20 cm from the scout, avoiding control on days after rainfall and when the soil is still damp. This stage was carried out in April 2015.

Figura 14   - Attack by leaf-cutting ants on eucalyptus leaves.
Source: Schumacher, 2010.

It is interesting to note that six months after planting the trees, anthills were found inside the planted area (Figure 15), but without significant damage from defoliation or attack. This observation and monitoring is important because, if the incidence of anthills intensifies and there is apparent damage caused by the reduction in leaf area, there may be a need for new control with formicide bait, now inside the plantation.

Figura 15 - Anthills in Neossolo in an area with eucalyptus forestry, Maçambará, RS. Source: Dick, 2017.

The next stage consisted of preparing the soil. As this was a new area, completely devoid of vegetation cover and with a gentle slope, it could have been done with minimal preparation, i.e. just opening the planting hole. However, due to the intense impact of raindrops, wetting and drying cycles, there was "undermining" of the surface layers of the soil. For this reason, scarification was carried out using an implement with a rod attached to a tractor, which turned the soil to a depth of 30 cm (Figure 16). This stage was carried out in May 2015.

Figura 16 - Neosol preparation in a new area for eucalyptus cultivation, Maçambará, RS. Source: Schumacher, 2015.

On the same day that the soil was prepared, liming and planting fertilisation were carried out. A total of 2 Mg ha^{-1} of dolomitic limestone and 150 kg ha^{-1} of K_2O were distributed, as well as 300 g plant^{-1} of triple superphosphate.

The seedlings were planted on the same day as the soil preparation, liming and pre-planting fertilisation. It was decided to plant the seedlings in May due to the reduction in wind speed and intensity during this period, which, in addition to wind erosion, can cause significant damage to the seedlings' stems through the abrasion of sand particles in the stem tissues. At the time of planting (Figure 17 a), the Eucalyptus urophylla clonal seedlings had an average height of 30 cm and were planted manually, using a manual forestry seedling planter (Figure 17 b), with a spacing of 3 m x 2 m between plants (initial density of 1,667 trees per hectare).

Figura 17 - a) Appearance of *Eucalyptus urophylla* seedlings after planting in Neossolo, Maçambará, RS. Source: Souza, 2015; b) Model of the manual forestry seedling planter used. Source: Agriprodutos, 2018.

There was no need for replanting because, twelve months after the forest was planted, the plant survival rate was over 98 per cent. This excellent result can be attributed to observing the right planting period (May), reducing wind damage and any water deficit, together with ant control and, above all, in response to fertilisation.

Top dressing fertiliser was added manually using a manual forestry seedling planter, which also has a compartment for fertiliser (Figure 18).

Figura 18   - Fertiliser application in side pits. Source: Souza, 2015

Fertiliser dosages containing macronutrients (nitrogen, phosphorus and potassium) were added in lateral pits, divided into six periodic applications: at 30, 75, 120, 180, 300 and 420 days after planting. In total, 156, 154 and 170 kg ha^{-1} of N, P2O5 and K2O were incorporated into the neosol, respectively (SOUZA, 2017). Fertilisation can be considered an important stage in environmental recovery, however, the need to add nutrients to forest plantations decreases after the canopy closes and litter production begins, due to nutrient cycling that can provide the nutritional supply for vegetation development (POGGIANI and SCHUMACHER, 2005).

$^{-1}$A total of 108 g of micronutrients were also applied manually and in instalments at 180, 300 and 420 days after planting, in the form of FTE compost (Calcium: 7.1%;

Sulphur: 5.7%; Boron: 1.8%; Copper: 0.8%; Manganese: 2.0%; Molybdenum: 0.1%; Zinc: 9.0%) (SOUZA, 2017). Especially in sandy soils, this addition of nutrients for the initial management of forest plantations is necessary due to the low fertility and rapid infiltration and percolation of water, which intensifies leaching and nutritional depletion in these soils (SANTANA et al., 2008).

As for the growth and productivity of this plantation, 24 months after planting, the average diameter at breast height (DBH) was 9.0 cm, the height was 9.6 m, the basal area was 10.5 m^2 ha^{-1} and the **wood volume was 56.5 m^3 ha^{-1}** (SOUZA, 2017). These results exceeded expectations and the productivity values are on a par with, or even higher than, those observed in site conditions more favourable to cultivation (DICK et al., 2016).

Based on these results, among the potential strategies for environmental recovery and promoting the ecosystem functionality of sandy soils, planting exotic tree species can be recommended. The practice of forestry in these areas aims to contain the advance of erosion processes, improve soil fertility, which is promoted through the accumulation of organic material (leaf litter) and soften the surface thermal oscillations of the Neossolo. In addition to these benefits, there is also a reduction in the speed of water infiltration and nutrient leaching, caused by root interception, the promotion of favourable conditions (shelter, food, humidity) for communities of edaphic organisms, among many other ecological features.

In addition to the efforts aimed at recovering soil quality and the functionality of ecological aspects, there are important productivity and income issues arising from the successful change in land use and cover (Figure 19). This new perspective on cultivation and land use promotes the reinsertion of sandy areas into the production chain of the Pampa biome, also taking into account the social aspect by encouraging producers to remain in the countryside.

Figura 19 - Landscape before and after eucalyptus forestry in Neossolo, Maçambará, RS. Source: Schumacher, 2015; Souza, 2017.

From twelve months after planting, the processes of litter production and accumulation were triggered. As well as being an ecological indicator for assessing the restoration of degraded areas, litter plays an important role in restoring soil fertility, as it provides nutrients through the decomposition and mineralisation of organic matter, formed by the accumulation of leaves, branches, bark and other plant material deposited on the forest floor (Figure 20) (ALONSO et al., 2015).

Figure 20 - Accumulated litter on the Neossolo in a eucalyptus plantation, Maçambará, RS. Source: Dick, 2017

Regarding the silviculture of exotic species grown in the Pampa biome, efforts and knowledge of improved techniques are still incipient and, as a result, the literature lacks records of studies that report on the ecological aspects and interactions of nutrient cycling and environmental recovery through the planting of eucalyptus in sandy soils.

CHAPTER 4

Silviculture in Cambissolo

The case study on eucalyptus forestry on Cambissolo was carried out in the central campaign region of the Pampa biome in Rio Grande do Sul, in the municipality of São Gabriel (Figure 21). In this area, trees have been planted for over ten years and the activities were monitored in the second rotation cycle, seven years after the first crop. Therefore, unlike the previous case where trees were planted in a new area, the practices described here were carried out in a renovated area where stumps and residues from the previous harvest remained on the ground.

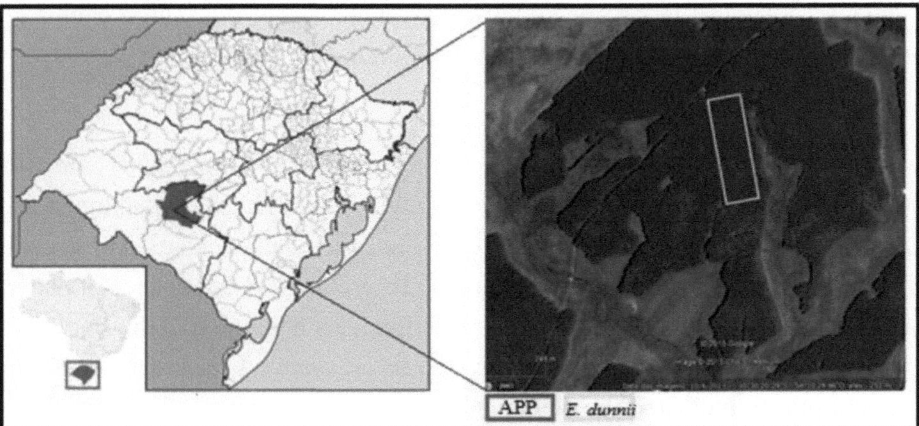

Figura 21 - Location of the eucalyptus plantation on Cambissolo, São Gabriel, RS. Source: Dick, 2018.

In the municipality of São Gabriel, livestock farming is significant and land use is dominated by fields, native and/or improved with the planting of annual and perennial pastures (Figure 22). Although the tradition of forestry is not so recent and consolidated in the region, the acquisition of land and encouragement of the practice, provided by investments from companies in the forestry sector, have changed the landscape of large areas of the municipality's rural environment. Planting trees in this new frontier may have been initially challenging, especially due to soil issues, as the Cambissolos found in this region are mostly medium-deep sandy soils with low fertility, originating from the geological formation of the Botucatu sandstone.

Figura 22 - Typical landscape and phytophysiognomy of the central campaign region of Rio Grande do Sul, São Gabriel. Source: MF Rural, 2018.

Before the land use was converted to forestry, the case study site was intensively used for livestock farming and, after the activity was abandoned and parts of the area were isolated, the initial colonisation and occupation of the soil with various species of grass could be seen. The ecological dominance of hardy, pioneering and aggressive plants, such as individuals belonging to the Baccharis genus, and intense biological invasion by the exotic species Eragrostis plana Nees (Annoni grass), indicated the degradation of the native grassland (DICK et al., 2016a). This may have occurred mainly due to the intense propagation and dispersal of annoni grass, which is difficult to control and contain, and the intensive livestock farming practised over long periods.

This situation is recurrent in the rural areas of the municipality of São Gabriel and often leads to the devaluation of the land, abandonment of livestock farming and rural exodus. However, there is viability in practising forestry, not least as an alternative source of income and recovery of soil quality, because, in addition to the type of soil, the climate in the region is favourable to tree growth, as long as the choice of frost-resistant plant material is observed.

The climate in São Gabriel is of the humid subtropical Cfa type (ALVARES et al., 2014), with a hot summer and no dry season, where the average temperature of the hottest month is over 22 °C and, in the coldest month, with many occurrences of frost, the averages vary between -3 °C and 18 °C. The lowest temperatures occur in June, July and August (winter season) and the highest in the summer period, from November to February. As for rainfall, the minimum in the driest month is over 40 mm and the average annual rainfall can vary from 1,600 to 1,900 mm per year[-1] , well distributed throughout the year (Figure 23).

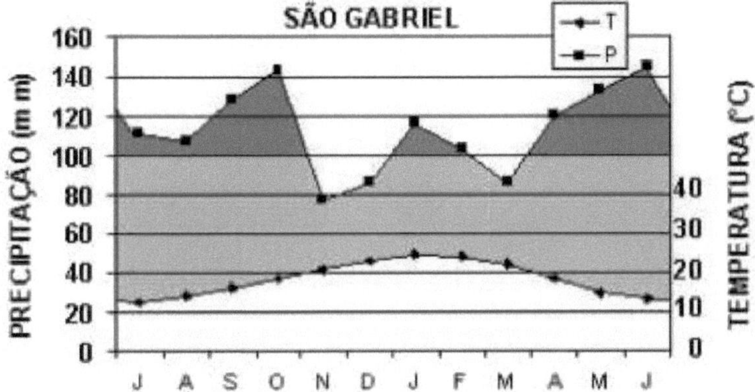

Figura 23 - Climate diagram for the municipality of São Gabriel, RS. Source: Buriol et al., 2007

In addition to the favourable climatic conditions, the morphological, chemical and physical analyses of the Cambissolo (Figure 24), a key stage in the practice of forestry, show that the soil has an average depth of over 50 cm, making it viable for growing trees. This soil was classified as typical Cambissolo Háplico Distrófico (EMBRAPA, 2013), with a sequence of A-Bi-C horizons in the profile, where the Bi horizon is of the incipient B type, i.e. it is in formation, but has developed enough colour and structure to be distinguished from the other horizons, and may contain rock fragments (STRECK et al., 2008).

Figura 24   - Cambissolo profile and sample collection for physical and chemical analysis. Source: Dick, 2018.

The texture of this Cambissolo is medium and the density values are within the appropriate range for sandy soil (ranging from 1.45 to 1.58 g cm^{-3}), with a predominance of gravel and pebbles in the granulometric composition. The fertility analysis showed that the organic matter content was low, as was the pH and cations such as Mg and Ca. The soil analyses show the residual effects of the fertilisation and liming carried out during the first crop rotation, with higher levels of P and K (DICK, 2018).

The dynamics of the implementation stages in this case study are different because of this first tree plantation in the area, which consisted of a stand of Eucalyptus saligna, planted in 2007, which was clear-cut at seven years of age (Figure 25). From this harvest, the wood with bark was removed from the site and all the stumps and residues (leaves and branches) were kept on the ground, partially distributed and girdled (DICK, 2018). Precautions should be taken regarding the bundling and distribution of residues because, when in large quantities and poorly disposed of, keeping this plant material on the ground can be a burden and hinder silvicultural processes.

Figura 25 - Distribution and girdling of residues after harvesting the first rotation of eucalyptus in Cambissolo, São Gabriel, RS. Source: Dick, 2014.

After characterising the soil, the species to be grown in the area was chosen, based mainly on climatic conditions and the purpose of the wood (pulp production), and the decision was made to plant Eucalyptus dunnii. The cultivation of this species is justified due to its good adaptation and tolerance to frost (HIGA et el., 2000) and water deficit (IPEF, 2015), its ability to utilise soil resources efficiently (GLENCROSS et al., 2014), as well as its considerably significant survival rates once established in the field (FILHO and SANTOS, 2005).

The silvicultural and operational procedures adopted for forestry implementation in this reform area consisted of: leaf-cutter ant control, soil preparation, weed control, liming and planting fertilisation, planting seedlings, sprouting control and top dressing fertilisation.

The leaf-cutting ant control was carried out in the entire area where there is effective planting and also used a bulk formicide bait with the active ingredient sulphuramide, applied close to the anthills 30 days before planting the seedlings. This stage, carried out in April 2014, was done manually, requiring the same care as mentioned in the description of the Neossolo (air and soil humidity, distance of the scout from the anthill, among other precautions).

The soil was prepared with a tractor fitted with a three-axle subsoiler, to an average depth of 50 cm, between the rows of the previous planting, i.e. between the stumps left over from the first cut, 30 days before the new planting (Figure

26). In this case study, the effective depth of preparation and the area of disturbance were greater when compared to the Neossolo, due to the existence of the root system from the previous cycle, which can cause physical impediments to the development of the new seedlings

of Eucalyptus dunnii. Preparation with three stems is justified by the arrangement of the residues on the ground, a method that makes it easier to clean the planting line, while at the same time incorporating fragments of residue into the soil. It should be emphasised that, despite the difficulties and care required in preparing the soil in this case of reform, maintaining the stump is extremely positive in the context of conservation forestry management. As well as minimising soil movement and exposure to erosive processes (mainly water and wind), these stumps (and roots) will be decomposed over time, becoming a source of nutrients and organic matter for the Cambissolo.

Figura 26   - Preparation of Cambissolo in a reform area for eucalyptus planting, São Gabriel, RS. Source: Dick, 2018.

To control weeds 20 days before planting the seedlings, 3 kg ha^{-1} of SCOUT® (glyphosate) herbicide was used in the total area and 1.2 L ha^{-1} of GOAL® (oxyfluorfen) in pre-emergence, in the planting line.

The next stage consisted of liming and planting fertiliser. Liming, carried out 30 days before planting, is necessary due to the primary need to make calcium and magnesium available to the plants, so 2 Mg ha^{-1} of dolomitic limestone was applied to the entire area. On the same day that the seedlings were planted, N-P2O5-K2O compost was applied manually to the furrow, with a formulation of 10:27:10, in the amount of 400 kg ha^{-1} . These measures correspond to 40 kg ha^{-1} of N (20 g plant^{-1}); 47 kg ha^{-1} of P (24 g plant^{-1}) and 33 kg ha^{-1} of K (17 g plant).$^{-1}$

The seedlings were planted in May 2014, as it was noted that in the region this is also the period of milder temperatures and water availability, as in the previous case study. The Eucalyptus dunnii genetic material came from seeds and therefore had greater genotypic variability and, at the time of planting, the average height of the seedlings was 30 cm (Figure 27 a). Planting was carried out manually, using a forestry seedling planter (as shown in Figure 17 b), with a spacing of 3.5 m x 2.15 m between trees (initial density of 1,961 plants per hectare). Despite the incidence of frost, the seedlings developed well (Figure 27 b), and up to 14 months after planting, the failure rate was 1.2%, with no need for replanting.

Figura 27   - a) Appearance of the seedlings shortly after planting; b) Appearance of the seedlings three months after planting in Cambissolo, São Gabriel, RS. Source: Dick, 2014.

In this case study, it is interesting to note that some of the factors that led to tree mortality were observed, such as: an excess of gnarled debris near the seedling, which ended up limiting the space and incidence of solar radiation to the plant; recurring problems with poor anchorage of the root system and growth of the tree trunk horizontally to the ground, due to lack of care and poor positioning of the seedling when planting was carried out.

For sprout control, one year after planting, 3 kg ha⁻¹ of SCOUT® (glyphosate) herbicide was applied to the entire area and 1.2 L ha⁻¹ of GOAL® (oxyfluorfen) was applied to the sprouts on the remaining stumps. Both weed and sprout control were carried out manually using knapsack sprayers. Special attention should be paid at these stages to the correct use of Personal Protective Equipment (PPE).

Top dressing fertiliser was applied in July 2015, 14 months after planting, by hand, while the fertiliser was spread on the soil taking into account the area of the crown projection (Figure 28). The amount of 294 kg ha⁻¹ (70 kg ha⁻¹ of N and 58 kg ha⁻¹ of K) was distributed over the soil, in a 24:00:24 fertiliser ratio and no phosphorus was applied as a top dressing, since this nutrient had already been added in the planting fertiliser.

Figura 28   - Top dressing fertilisation applied to the projection of the eucalyptus canopy, São Gabriel, RS. Source: Dick, 2015.

As for the growth and productivity of this plantation, from eight to 36 months after planting, the average diameter at breast height (DBH) increased by an average of 10 cm and the increase in height during this period was 14 m; while the volume went from 1.21 m³ ha⁻¹ to **107 m³ ha⁻¹** after three years of cultivation

(Figure 29). It is interesting to note that the total biomass production of the trees in this case study was higher than that observed in a plantation of the same species grown on Argissolo (DICK et al., 2016)

Figura 29 - Eucalyptus trees at eight and thirty-six months after planting in Cambissolo, São Gabriel, RS. Source: Dick, 2014 (a); Momolli, 2017 (b).

It should be noted that these significant productivity results can be attributed, in addition to the effect of fertilisation, to the maintenance of residues from the previous rotation on the soil, which also contributes to improving fertility, especially in the surface layer, and acts to reduce erosion.There is a tendency for the quality of the Cambissolo to improve as the rotation progresses because, during the process of decomposition and mineralisation of the remaining stump and cultural remains from the harvest, nutrients will be released from the organic matter into the soil, which will be absorbed by the eucalyptus plants (FARIA et al... 2009), 2009).

CHAPTER 5

Silviculture on Argissolo

The case study on the cultivation of eucalyptus trees on Argissolo was conducted in the western campaign region of the Pampa biome, in the municipality of Alegrete, Rio Grande do Sul (Figure 30). The expansion of forestry in this region, as in the previous study on Cambissolo, is also the result of investments by the forestry sector.

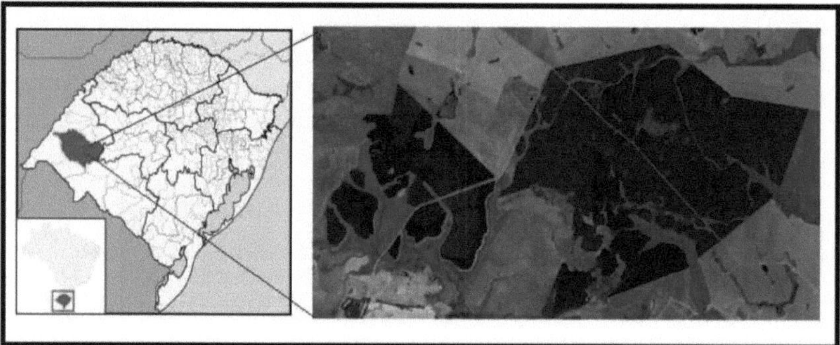

Figura 30 - Location of eucalyptus plantation on Argissolo, Alegrete, RS. Source: Wikipedia, 2018; Google Earth: 2018.

The scenario of economic projection through the installation of pulp mills in the region (which had not yet been consolidated), coupled with the abandonment of livestock farming due to the extensive areas of degraded pasture, especially through biological contamination with anonni grass (Figure 31), the devaluation of land prices and the lack of incentives for farmers, were some of the motivations driving the advance of forestry in this new frontier.

Figura 31 - Landscape of Pampa biome grasslands and forestry in the western campaign region, Alegrete, RS. Source: Dick, 2015.

The region's climatic conditions are also favourable for forestry, as long as the ecophysiology of the eucalyptus species and the planting site are observed (incidence of frost, soil humidity, effective depth, among others).The region's climate (Figure 32) is of the humid subtemperate type, with frosts in the winter season (July to September) and summers that can be dry, with an average annual temperature of 18.6 °C. The average annual rainfall is 1,575 mm, well distributed throughout the year. The average annual rainfall is 1,574 mm, which is well distributed throughout the year.

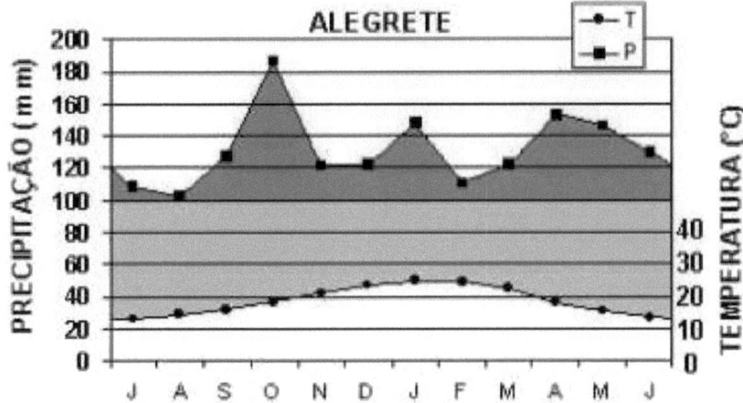

Figura 32 - Climate diagram for the municipality of Alegrete, RS. Source:

Buriol et al., 2007.

By opening a trench (Figure 33), it can be seen that the soil in the study area, which comes from the geological formation of the Botucatu sandstone, is morphologically classified as typical dystrophic red Argissolo (Figure 34). Of the soil classes assessed, the Argissolo is more weathered, deeper and has a textural B horizon, which often, when too clayey and compacted, can hinder the development of the root system, reducing water percolation and the volume of soil to be explored by the roots.The physical analysis carried out before planting the trees revealed that the texture is sandy, varying from sandy loam to sandy loam, with a depth varying from medium to high. The fertility analysis described in Corrêa et al. (2013) found low organic matter content (1.0%) and very low pH (4.7). The Ca and Mg contents were also low, and the K and P contents were very low. The effective CEC is average, as well as very low base saturation and high Al saturation, revealing the soil's low fertility.

Figura 33   - Aspect of the trench for collecting samples in Argissolo, Alegrete, RS. Source: Schumacher, 2012.

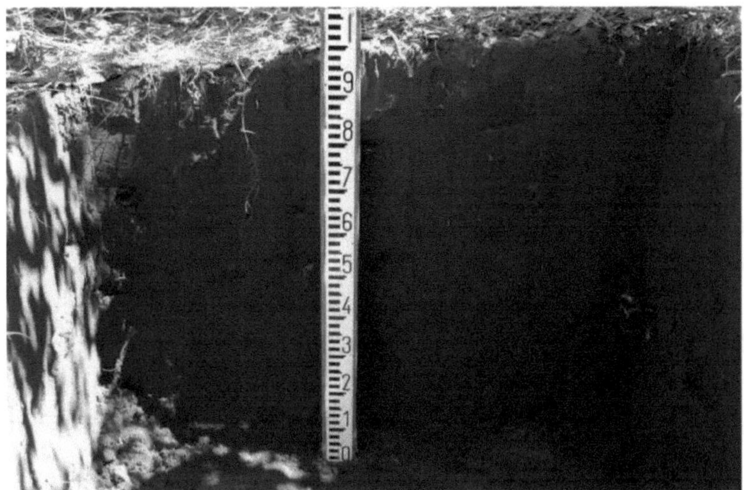

Figura 34   - Argissolo profile in an area with eucalyptus cultivation, Alegrete, RS. Source: Schumacher, 2010.

Forestry was implemented on this Argissolo in the new area system, i.e. there had been no previous tree cultivation, but rather a degraded native field with a high incidence of exotic grasses. Therefore, unlike the previous case, there is no waste on the ground, but rather creeping plants, facilitating the initial stages of preparation and cultivation.

Eucalyptus dunnii was chosen as the species to be planted in the area, considering the same conditions of resistance to the frequent incidence of frost in the region and the purpose of the product (cellulose), as discussed in the study on Cambissolo.The silvicultural operations in this case study consisted of: leaf-cutting ant control, soil preparation, phosphating, planting fertiliser, planting seedlings with irrigation, weed control and top dressing fertilisation.As in the other studies, leaf-cutting ant control in this case was also carried out 30 days before planting, manually with the application of bulk formicide baits, with the active ingredient sulphuramide, in and around the actual planting area. It should again be noted that the application of manual techniques at this stage is one of the conditions for obtaining an environmental licence, which is required for the practice of forestry.

As for soil preparation, this was carried out in a new area with previous field occupation, by subsoiling, using a tyre tractor, at an average depth of 60 cm of soil revolvement, in October 2008. This greater depth of pre-soil, when compared to the case of the Neossolo for example, is justified by the presence of the textural B horizon in the soil profile. Along with this stage, on the same day, phosphating was applied to the planting line, with 300 kg ha^{-1} of Gfsa natural phosphate, incorporated up to a depth of 20 cm. After this stage, approximately 15 days after preparing the soil, planting fertiliser was applied, with 300 kg ha^{-1} of NPK, in the 06-30-06 + 0.6 % B formulation, in lateral pits, with the help of the forestry seedling planter (Figure 17 b).The eucalyptus seedlings were planted in November 2008, so irrigation was necessary due to the higher temperatures and lower rainfall during this period. The seedlings, which came from seeds with an average height of 25 cm, were planted using a manual planter (Figure 35), using a spacing of 2.0 m x 3.5 m between plants, with an initial density of 1,428 trees per hectare.

Figura 35   - Manual planting of eucalyptus seedlings in Argissolo, Alegrete, RS. Source: Souza, 2008.

Irrigation was carried out on the same day as planting using a tractor and a water tank with hoses, which accompanied the operators to the planting site, providing water as soon as the seedlings were placed in the ground (Figure 36).

Figura 36 - Irrigation of eucalyptus seedlings in Argissolo, Alegrete, RS, Source: Souza, 2008.

Manual weeding methods were used to control weeds, with a hoe, in the planting row and between the rows, and the application of herbicide (glyphosate) between the rows. This control is more intense in the first few months after planting and reduces as the trees grow, as the canopy closes and the soil is shaded, reducing the incidence of invasive plants, which compete with the eucalyptus trees for nutrients, water and solar radiation. As this is a new area, with field occupation, there was a need for frequent control interventions.

The top dressing stage began 90 days after planting, when 140 kg ha^{-1} of NPK was applied, using the 22-05-20 formulation, plus 0.2% B and 0.4% Zn. The third fertilisation took place 270 days after planting, using 140 kg ha^{-1} of NPK, 22-00-18, enriched with 1.0% S and 0.3% B. These fertilisations were applied mechanically between the rows.

At 60 months after planting Eucalyptus dunnii, the average diameter at breast height (DBH) was 12 cm, height 13 m, volume with bark **124.3 m^3 ha^{-1}** and total biomass production 58.76 Mg ha^{-1} (DICK et al., 2016). Figure 37 shows the appearance of the plantation at the end of the rotation, 84 months after planting, and the architecture and root development in Argissolo, where a considerable volume of soil was exploited.

Figura 37   - Aspect of eucalyptus plantation grown on Argissolo at 84 months (a) and root system at 60 months after planting (b). Source: Dick, 2015 (a); Schumacher, 2010 (b).

The low volumetric development of this stand may have been influenced by low soil fertility. It may also be linked to the unfavourable planting period (November), the influence of the genetic material (species and origin of the seedling), as well as the lower levels of fertilisation applied when compared, for example, to the case study of eucalyptus cultivation on Cambissolo. However, despite not achieving significant productivity, growing trees in this region is a viable economic alternative, as well as improving the chemical and physical properties of the Argissolo.

Especially in the Pampa region, where livestock farming is unfavoured due to the quality of the native grassland, there is a growing likelihood of abandonment of the activity and a rural exodus. Forestry as a development strategy may be feasible and promising in a scenario where extensive rural areas are compromised by biological invasion and loss of productive capacity.

CHAPTER 6

Final considerations

Forestry, like other activities that use natural resources, should always be practised in a conservationist way, with the aim of minimising environmental impacts. The adoption of reduced tillage, with tillage only in the planting line, was observed in all the case studies, and was feasible regardless of the soil class. Particularities inherent to these soil classes should always be considered when preparing the soil, especially when choosing the implement (shank, subsoiler, scarifier, harrow, etc.) and the effective depth of tillage. Observing these issues is essential because correct soil preparation is one of the important stages of operational forestry, which can lead to increased tree productivity at the end of the rotation or exacerbated losses of soil and nutrients through erosion.Silvicultural management is also strictly related to the history of land use. In the case studies, we saw different situations (new and renovation areas) and different operations depending on the presence of stumps, residues, sandy soil that was completely exposed or occupied by grasses. This shows that standardised and generalised techniques are not applied to forestry activities: in order to make operational decisions, each site must be assessed individually.

Considering the aspects of soil, climate, location and suitable species are key factors for the successful cultivation of trees in the Pampa biome, which bring not only environmental benefits but also economic and social perspectives. Valuing forestry as a tool for inserting resources into rural property, linked to ecosystem services that improve soil quality, is fundamental in a scenario of over-exploitation of the countryside by livestock and conventional agriculture.

CHAPTER 7

Bibliographical references

ABRAF ■ BRAZILIAN ASSOCIATION OF PRODUCERS OF PLANTED FORESTS. **Statistical Yearbook 2013: base year 2012.** Available at: http://www.abraflor.org.br. Accessed on: 29/09/2017.

AGEFLOR - GAUCHO ASSOCIATION OF FORESTRY COMPANIES. **The forest-based industry in RS:** data and facts - 2014 base year. Porto Alegre, RS: AGEFLOR, 2015. 40p.

AGNELI, A. Guidelines for choosing Eucalyptus species. **IPEF,** 2005. Available at: http//www.ipef.br/identificacao/eucalyptus/indicacoes.asp. Accessed on: 21 Mar 2017.

ALONSO, J.M. et al. Litter transport in forest restoration plantations at different spacings. **Revista Ciência Florestal**, Santa Maria, v. 25, n. 1, p. 1-11, 2015.

ALVARES, C.A. et al. Kõppen's climate classification map for Brazil. **Meteorologische Zeitschrift** v.22, n.6, p :711-728, 2014.

BARROS, N.F.; NEVES, J.C.L.; NOVAIS, R.F. Nutrition and mineral fertilisation of Eucalyptus. In: VALE, A.B. et al. **Eucalyptus cultivation in Brazil: silviculture, management and ambience.** Editora UFV, 1ed, Viçosa: UFV, 2014. p: 187-207.

BEGON, M.; TOWNSEND, C.R.; HARPER, J.L. **Ecology: from individuals to ecosystems.** Editora Artmed, 4ed. Porto Alegre, 2007, 752 p.

BELDINI, T.P. et al. The effect of plantation silviculture on soil organic matter and particle-size fractions in Amazonia. **Revista Brasileira de Ciência do Solo.** v.33, p:1593- 1602, 2009.

BINKLEY, D. et al. The interactions of climate, spacing and genetics on clonal Eucalyptus plantations across Brazil and Uruguay. **Forest Ecology and Management,** v.405, p: 271283, 2017.

BOLDRINI, I. I. et al. **Pampa Biome:** floristic and physiognomic diversity. Editora Pallotti, Porto Alegre, RS, 2010. 64p.

BOOTH, T.H. et al. Native forests and climate change: Lessons from eucalypts. **Forest Ecology and Management.** v.347, p: 18-29, 2015.

BOOTH, T.H. Eucalypt plantations and climate change. **Forest Ecology and Management,** v301, p: 28-34, 2013.

BURIOL, G.A. et al. Climate and natural vegetation of the state of Rio Grande do Sul according to the Walter and Lieth climate diagram. **Ciência Florestal**, Santa Maria, v. 17, n. 2, p. 91100, Apr-Jun, 2007.

CORRÊA, R. S.; SCHUMACHER, M. V.; MOMOLLI, D. R. Deposition of litter and macronutrients in a stand of Eucalyptus dunnii Maiden on degraded natural pasture in the Pampa Biome. **Scientia Forestalis** (IPEF), v. 41, p. 65-74, 2013.

DICK, G. **Mineral fertilisation in** Eucalyptus dunnii **Maiden: effects on nutrient stocks.** PhD thesis, Postgraduate Programme in Forestry Engineering, Federal University of Santa Maria, RS. 96p.

2018.

DICK et al. Strategies for the restoration of permanent preservation areas in the Pampa biome. In: I International Congress of the Pampa and III Seminar on Sustainability in the Campanha Region. **Proceedings...** Universidade Federal de Santa Maria, Santa Maria, RS, 2016.a Available at: http://coral.ufsm.br/cipa/index.php/anais. Accessed on: 10/04/2017.

DICK, G. et al. Nutrient balance in the soil-plant system in a Eucalyptus dunnii Maiden stand in the Pampa biome, RS. In: III Brazilian Eucalyptus Congress. **Proceedings...** Vitória, ES. 2015.

DICK, G. et al. Quantification of biomass and nutrients in a Eucalyptus dunnii Maiden stand established in the Pampa biome. **Forest Ecology and Nutrition.** v.4, n.1, p: 01-09, 2016.

EMBRAPA (Brazilian Agricultural Research Corporation). **Brazilian soil classification system.** Embrapa Solos, 3ed. Rio de Janeiro/RJ, 2013. 306p.

EMBRAPA (Brazilian Agricultural Research Corporation). **Manual of soil analysis methods.** National Soil Research Centre 2ed. Rio de Janeiro, 1997.

FAO - FOOD AND AGRICULTURE ORGANISATION OF THE UNITED NATIONS. **Global Forest Resources Assessment 2015**: How are the world's forests changing? 2ed. Rome: FAO, 2016. 54p.

FAO - FOOD AND AGRICULTURE ORGANISATION OF THE UNITED NATIONS. **Status of the World's Soil Resources:** Main report Prepared by Intergovernmental Technical Panel on Soils (ITPS) Rome, 2015, 648p.

FARIA, G.E. et al. Soil chemical characteristics at different distances from the eucalyptus trunk and at different depths. **Revista Árvore.** v.33, n.5, p:799-810, 2009.

FILHO, E.; SANTOS, P.E.T. Considerations on planting Eucalyptus dunnii in the state of Paraná. **Technical Communication 141.** Colombo, PR. December 2005.

FLORES, T.B. et al. Eucalyptus **in Brazil: climate zoning and identification guide.** IPEF, Piracicaba, SP. 2016, 448p.

GLENCROSS, K. et al. Basal area increment is unaffected by thinning intensity in young Eucalyptus dunnii and Corymbia variegata plantations across different quality sites. **Forest Ecology and Management.** v.138, p: 326-333, 2014.

GONÇALVES, J.L.M. et al. Edaphoclimatic characterisation and soil management of areas with eucalyptus plantations. In: SCHUMACHER, M.V. VIERA, M. **Silvicultura do Eucalipto no Brasil.** Editora UFSM, Santa Maria. 2015, 308p.

GONÇALVES, J.L.M. et al. An evaluation of minimum and intensive soil preparation regarding fertility and tree nutrition. In: GONÇALVES, J.L.M.; BENEDETTI, V. **Forestry nutrition and fertilisation.** IPEF: Piracicaba, 2004. p. 13-64.

GONÇALVES, J.L.M. Soil conservation. In: Soil **conservation and cultivation for forest plantations.** IPEF - Forestry Research and Studies Institute - Piracicaba, 2002. 498p.

HIGA, R.C.V. et al. Frost resistance and resilience in Eucalyptus dunnii Maiden planted in Campo do Tenente, PR. **Boletim de Pesquisa Florestal.** n.40, p: 67-76, 2000.

IBÁ - BRAZILIAN TREE INDUSTRY. **Data and statistics 2017.** Available at:

<http://iba.org/pt/dados-e-estatisticas/cenarios-iba>. Accessed on: 15 January 2018.

IPEF - FORESTRY RESEARCH INSTITUTE. Forest Species Identification Key. Available at: http://www.ipef.br/identificacao/cief/especies/dunnii.asp. Accessed on: 12 December 2015.

LACLAU, J-P. et al. Organic residue mass at planting is an excellent predictor of tree growth in Eucalyptus plantations established on a sandy tropical soil. **Forest Ecology and Management**. v.260, p: 2148-2159, 2010.

LEITE, F. P. et al. Alterations of soil chemical properties by eucalyptus cultivation in five regions in the Rio Doce Valley. **Brazilian Journal of Soil Science**. v.34, p: 821-831, 2010.

LEPSCH, I. **Soil formation and conservation.** São Paulo: Oficina de Textos, 2002.

LIMA, W. P. **Environmental impact of eucalyptus.** Editora USP: São Paulo, 1996. 307p.

MEURER, E. J.; RHEINHEIMER, R.D.; BISSANI, C.A. Sorption phenomena in soils. In: MEURER, E. J. **Fundamentos de química do solo**. Editora Evangraf LTDA, Porto Alegre, 2010. p.107-147.

PAES, F.A.S.V. et al. Impact of harvest residue management, soil preparation and fertilisation on eucalyptus productivity. **Revista Brasileira de Ciência do Solo**. v.37, p: 1081-1090, 2013.

POGGIANI, F.; SCHUMACHER, M.V. Nutrient cycling in native forests. In: GONÇALVES, J.L.M.; BENEDETTI, V. **Nutrição e Fertilização Florestal**. IPEF - Piracicaba, 2005, 427p.

SANTANA, R. C. et al. Nutrient allocation in eucalyptus plantations in Brazil. **Revista Brasileira de Ciência do Solo**. v.32, p: 2723-2733, 2008.

SILVA, I.R; MENDONÇA. E. S. Soil organic matter. In: NOVAIS, R. F.et al. eds. **Soil Fertility**. 1ª ed. Viçosa/MG, Sociedade Brasileira de Ciência do Solo, 2007. p.275-374.

SOUZA, H.P. **Silvicultural aspects of** Eucalyptus urophylla **s.t. Blake under different fertilisation regimes in sandy soil in the Pampa biome.** Doctoral thesis -

Postgraduate Programme in Forestry Engineering, Federal University of Santa Maria, UFSM, 91 p. 2017.

SUERTEGARAY, D.M.A. Erosion in the Southern Fields: Arenisation in Southwestern Rio Grande do Sul. **Brazilian Journal of Geomorphology**, v. 12, n. 3, 2011.

STRECK, E.V. et al. **Soils of Rio Grande do Sul.** 2.ed. Porto Alegre/RS, EMATER/RS- ASCAR, 2008. 222p.

TAIZ, L.; ZEIGER, E. **Plant Physiology.** Editora Artmed, 5ed. Porto Alegre, 2013.

TEDESCO, M.J. et al. **Analysis of soil, plants and other materials.** 2ed. Porto Alegre, UFRGS, 1995. 174p.

VERDUM, R. STRECK, E.V.; VIEIRA, L.F.S. Soil degradation in Rio Grande do Sul. In: GUERRA, A.J.T.; JORGE, M.C.O. **Degradação dos solos no Brasil**. Editora Bertrand Brasil, Rio de Janeiro, 2014. p. 87-125.

Buy your books fast and straightforward online - at one of world's fastest growing online book stores! Environmentally sound due to Print-on-Demand technologies.

Buy your books online at
www.morebooks.shop

Kaufen Sie Ihre Bücher schnell und unkompliziert online – auf einer der am schnellsten wachsenden Buchhandelsplattformen weltweit! Dank Print-On-Demand umwelt- und ressourcenschonend produziert.

Bücher schneller online kaufen
www.morebooks.shop

Printed by Books on Demand GmbH, Norderstedt / Germany